THEMATIC UNIT

# Farm

***Written by Cynthia Holzschuher***

*Edited by Dorinda Mas, M.S. Ed.*

*Illustrated by Barb Lorseyedi*

*Cover Art by Ken Tunell*

***Teacher Created Materials, Inc.***
P.O. Box 1040
Huntington Beach, CA 92647

Made in U.S.A.

**ISBN 1-57690-059-2**

# Table of Contents

# Introduction

The **Farm** contains three units on literature selections, as well as activities that extend across the curriculum to language arts, math, social studies, music, art, and life skills. The activities are designed to address all learning modalities. Directions for student-made books, singing, games, crafts, planting seeds, and a culminating activity that allow students to synthesize their knowledge and produce products which can be shared beyond the classroom, are all contained in this unit.

This thematic unit includes:

- ❑ **literature selections**—summaries of three children's books with related lessons and skill pages that cross the curriculum
- ❑ **poetry**—original poetry and suggestions enabling students to write and publish their own poems
- ❑ **language experience and writing ideas**—a student-made book related to each literature selection
- ❑ **bulletin board ideas**—suggestions and plans for interactive boards
- ❑ **curriculum connections**—activities covering a wide range of skills
- ❑ **culminating activity**—shared activity where students synthesize learning
- ❑ **bibliography**—literature, nonfiction books, audio-visual aids, and computer software related to the theme

To keep this valuable resource intact so it can be used year after year, you may wish to punch holes in the pages and store them in a three-ring binder.

# Introduction *(cont.)*

## Why an Integrated Approach?

The strength of an integrated approach is that it involves children in using all modalities of communication: reading, writing, speaking, listening, illustrating, and interacting. Communication skills are interconnected and integrated into lessons that emphasize the whole of language. Balancing this approach is our knowledge that every whole, including individual words, is composed of parts, and directed study of those parts can help a student to master the whole. Experience and research tell us that regular attention to phonics, other word attack skills, spelling, etc., develops reading mastery, thereby completing the unity of the language experience. The child is thus led to read, write, spell, speak, and listen confidently in response to a literature experience introduced by the teacher. In these ways, language skills grow rapidly, stimulated by direct practice, involvement, and interest in the topic at hand.

## Why Include Whole Language?

A whole language approach involves children in using all modes of communication: reading, writing, listening, observing, illustrating, experiencing, and interacting. Communication skills are interconnected and integrated into lessons that emphasize the whole of language rather than isolating its parts. The lessons revolve around selected literature. For example, reading is not taught as a separate subject from writing and spelling. A child reads, writes, spells appropriately for his/her level, speaks, listens, etc., in response to a literature experience introduced by the teacher. In this way, language skills grow naturally, stimulated by involvement and interest in the topic at hand.

## Why Thematic Planning ?

An integrated program is best implemented with thematic planning. The teacher plans classroom activities correlated to specific literature selections centering around a predetermined theme. Students tend to learn and retain more when they are applying their skills in an interesting and meaningful context. Both teachers and students will be freed from a day that is broken into unrelated segments of isolated drill and practice.

## Why Cooperative Learning?

Students need to learn social appropriateness as well as academic skills. Group activities are part of living, and it is important to consider social objectives in your planning. Students working together will select leaders and designate responsibilities within their groups. The teacher is present to explain social goals and monitor interaction.

## Why Make Books?

Groups of children may produce a book as a cooperative learning project. In so doing they gain experience in reading, writing, spelling, illustrating, as well as building self-esteem. Student-made books in this unit help teach a skill or recall a story. The books may be shared with friends and family members.

# *Family Farm*

## *by Thomas Locker*

### Summary

*The family is in danger of losing their farm because they are not making enough money from the sale of milk and corn. Father takes a job in a factory, and Mother and the children contribute by raising and selling pumpkins and flowers.*

### Concepts:

- The food we eat comes from a farm.
- The farmer and his family have many jobs.
- A farm raises crops and animals.

### Concept Activities:

- Let students cut out magazine pictures of food and glue them to posters labeled fruit, vegetable, meat, grain.
- With the children make a list of the jobs of the farmer in the book.
- Role play activities, such as planting seeds, feeding chickens, fixing a tractor, and selling pumpkins and flowers at a roadside stand. Copy this chart onto the board. Let the class help finish this list of family members' chores.

| Father | Mother | Grandpa | Sarah and Mike |
|---|---|---|---|
| factory work | plant flowers | milk cow | feed pigs |
| milk cows | plant pumpkins | clean barn | work at farm stand |

- If you live in or near a farming community, list the animals and crops that are grown there. Find pictures of them and show them to the students. Try to arrange a visit to a local farm to see live animals.
- Use the vocabulary word and picture cards on pages 8 and 9 to make sentences that emphasize these concepts or for matching and writing activities.

### Additional Activities:

1. Determine with your students the setting for this story. In what state might it take place? How do you know? Also discuss the need to raise money to keep the family farm. At the end of the story, each problem will have a clear solution. Discuss the solutions to the problems with your students. If appropriate have them write or illustrate the problems and their solutions.
2. Invite a farmer to visit your classroom to talk about his or her work. Encourage children to ask about how a tractor is used. Ask what is the best/worst thing about living on a farm.

# Family Farm *(cont.)*

## Additional Activities *(cont.)*

3. Make a list of farm products for each letter of the alphabet. Ask each child to draw a farm product along with its letter on a six-inch (15 cm) square of paper. Glue the pictures in alphabetical order to a large paper. Cut around the border to make an ABC quilt of farm products.
4. Ask children if they have ever attended a county fair. What kinds of products and activities did they see? Discuss what was expected from Sarah and Mike in raising their calf, Derinda. Do you think Derinda will be a winner? Have children explain their answers.
5. Complete the Farm Riddle Book on page 7 with the children.
6. Read the poem "Farm Day" on page 11. Have the children repeat the animal sounds.
7. Let students complete the work sheet on page 12, matching the animals and crops.
8. Show the pictures in the book of the farm family's roadside stand. Ask your students if they have ever shopped at such a place. Compare and contrast that experience with shopping at a supermarket. Distribute the work sheet on page 13 for practice reading numerals and drawing sets.
9. Give children an opportunity to help the farmer find his way through the pumpkin patch maze on page 16.
10. Prepare the matching game on page 10. The game pieces may be used for matching like pictures, words and pictures, or sets and numerals.
11. What jobs do the children do to help on the farm? What job would be the most fun? Would your students like to live on a farm? Why? Why not?
12. Complete a chart with information from the story. Example: Plants and Animals.
13. Explain that plants need sun, water, and rich soil to grow. What would happen to a farmer's crops if there was no rain for a long time? No sunshine? Poor soil? Complete the work sheet on page 14.
14. Invite your students to save seeds from foods they eat at home, dry them on paper toweling, and bring them to share with the class. Glue each different kind of seed to a card, showing the name and a picture of the food that the seed produces.

## Culminating Activities:

1. Use the directions on pages 17 and 18 to plant several kinds of farm products. Make picture charts to show the growth of each plant. Follow them for several weeks.
2. Sing *"Old McDonald Had a Farm"* and *"The Farmer in the Dell"* on page 78.

# Farm Riddle Book

Draw the answer to each riddle. Cut apart the pages and staple into a book.

| | | |
|---|---|---|
| Name_______________ <br><br> I am yellow. I am used in cereal, bread, or eaten on the cob. What am I? 1. | I am a fruit. I can be red, yellow, or green. I grow on trees. What am I? 2. | I am a big animal. I give milk. I can be black or brown. What am I? 3. |
| I am a small animal. I have feathers. I lay eggs in a nest. What am I? 4. | I help the farmer plow the fields, plant seeds, and harvest the crops. I am a machine. What am I? 5. | I am a farm building where animals live and hay is stored. What am I? 6. |

# Vocabulary Word and Picture Cards

**To the teacher:** Copy the vocabulary cards on heavy paper. Color the pictures. Laminate for durability. Cut apart pictures and words. Use for matching, spelling, or oral language activities.

**Word List:**

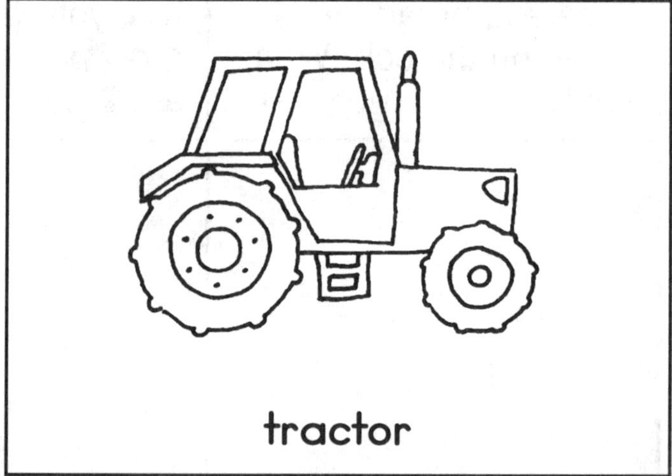

# Vocabulary Word and Picture Cards *(cont.)*

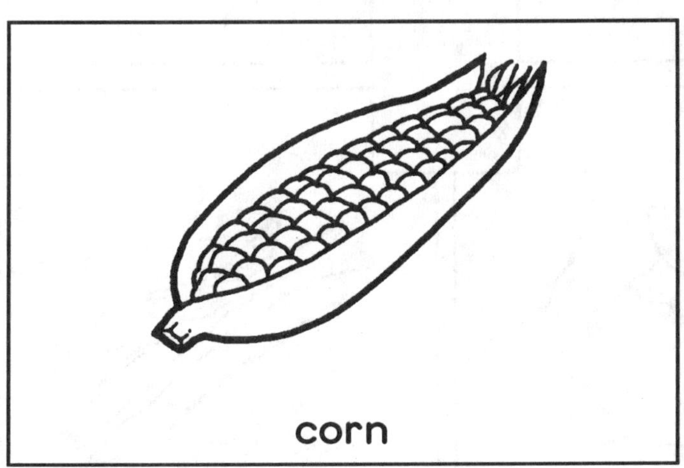

# Farm Field Matching Game

This game has been designed for matching words and pictures or sets and numerals.

**You will need:**

- two sets of pattern cards
- tagboard
- crayons or markers
- glue
- sharp scissors, craft knife
- clear shipping tape
- one sheet green construction paper 9" x 12" (23 cm x 30 cm)
- a sturdy cardboard box about 9" x 12" (23 cm x 30 cm)

**Directions:**

1. Copy two sets of the pattern cards. Determine the skill for your game. Add words and pictures or numerals and sets to the bottoms of the cards or on the tab. Color the pictures and glue them to tagboard. Laminate.
2. Glue the green construction paper to the top of your box. Flatten the box lid and laminate. Tape the lid tightly to the box bottom. With the craft knife, cut in the box lid 12 slots big enough to hold the game pieces.
3. Students will take turns choosing two cards from the board. If the bottoms match, the cards are removed. If there is no match, cards are replaced in their original spaces. The student with the most matches at the end of the game is the winner.

# Farm Poems

### Farm Day

**Teacher:** Animals living in the barn,
Wake up early on the farm.
**Class:** The cows say, "Moo,"
The pigs, "Oink, Oink,"
And the rooster goes, "Cock-a-doo."
**Teacher:** The farmer and his family too,
All have chores that they must do.
**Class:** The cows say, "Moo,"
The pigs, "Oink, Oink,"
And the rooster goes "Cock-a-doo".
**Teacher:** They milk the cows and feed the hen,
And put the pigs out in their pen.
**Class:** The cows say, "Moo,"
The pigs, "Oink, Oink,"
And the rooster goes, "Cock-a-doo."
**Teacher:** A farmer's work is never done,
From early morn 'til setting sun.
*(Children may memorize the repeated section.)*

### The Scarecrow

Scarecrow standing in the field
On a bright and sunny day,
Don't forget to do your job.
Scare the hungry crows away!

### The Farmer

Sun is coming up
Farmer's out the door,
He will go to milk the cows,
And start his daily chores.

Sun is going down
Horse is in the stable,
All the fields are planted now,
Supper's on the table.

### Suggestions for Using the Poems

These techniques help to promote reading readiness and introduce children to the "idea" of reading as well as provide practice in oral language skills.

- You may print the poems on sentence strips in two different colors. One color indicates lines the teacher reads, and the other indicates the lines the class reads.
- Print the poem on poster board with magazine pictures of farm animals. Pictures may be cut from magazines.
- Make puppets out of paper lunch sacks and have several groups of children be the various animals.
- Practice tracking the words with your children while reading the poems.

Name ______________________________

# Match the Animals and Crops

1. Color, cut, and match the farm crops and animals to their products.
2. Glue at the tab so the products may be seen.
3. Lift the pictures of the farm crops and animals so you may see what products each make.

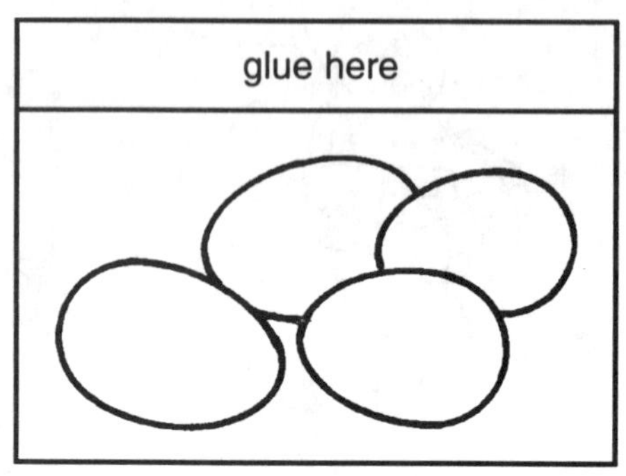

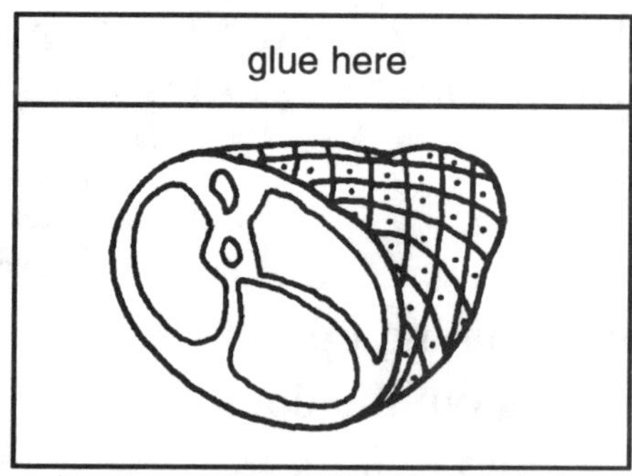

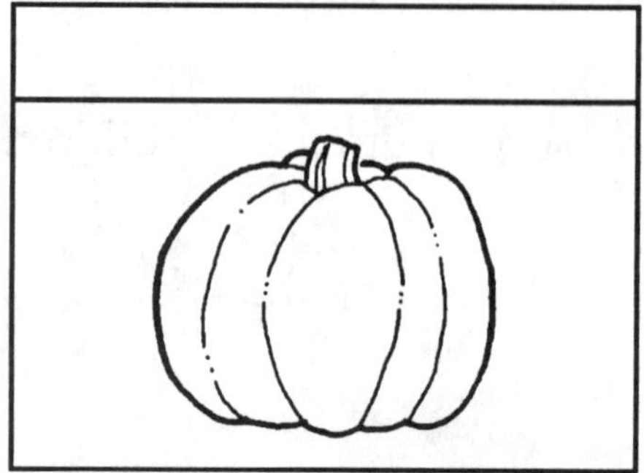

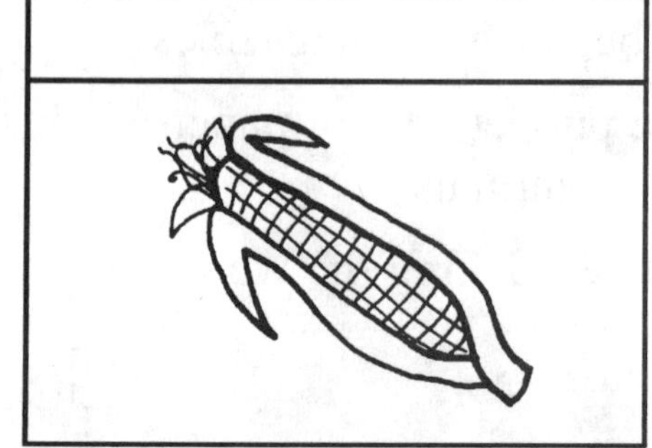

 

Name ______________________________

# Roadside Stand

Draw items found on a farm to show the number and color.

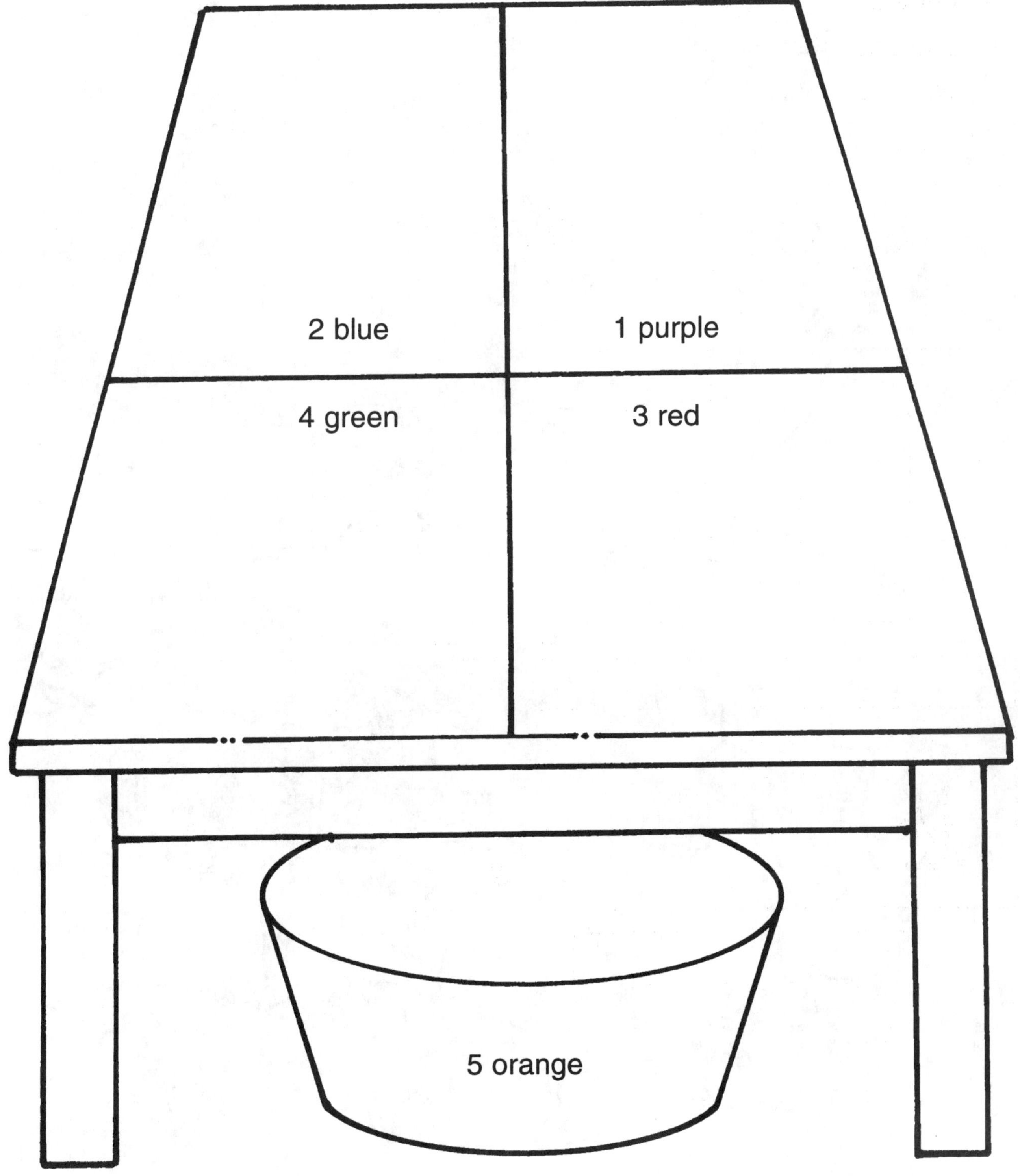

Name ______________________________

# What Do Plants Need to Grow?

Add water, sunshine, and rich soil to help the garden grow. Complete the picture and color it.

Name ______________________________

# Pumpkin to Jack-O'-Lantern

Cut and glue the jack-o'-lantern in the correct order.

(cut apart)

| | |
|---|---|
| 1. | 2. |
| 3. | 4. |

Name ________________________________

# Pumpkin Patch Maze

Help the farmer find his way back to the barn.

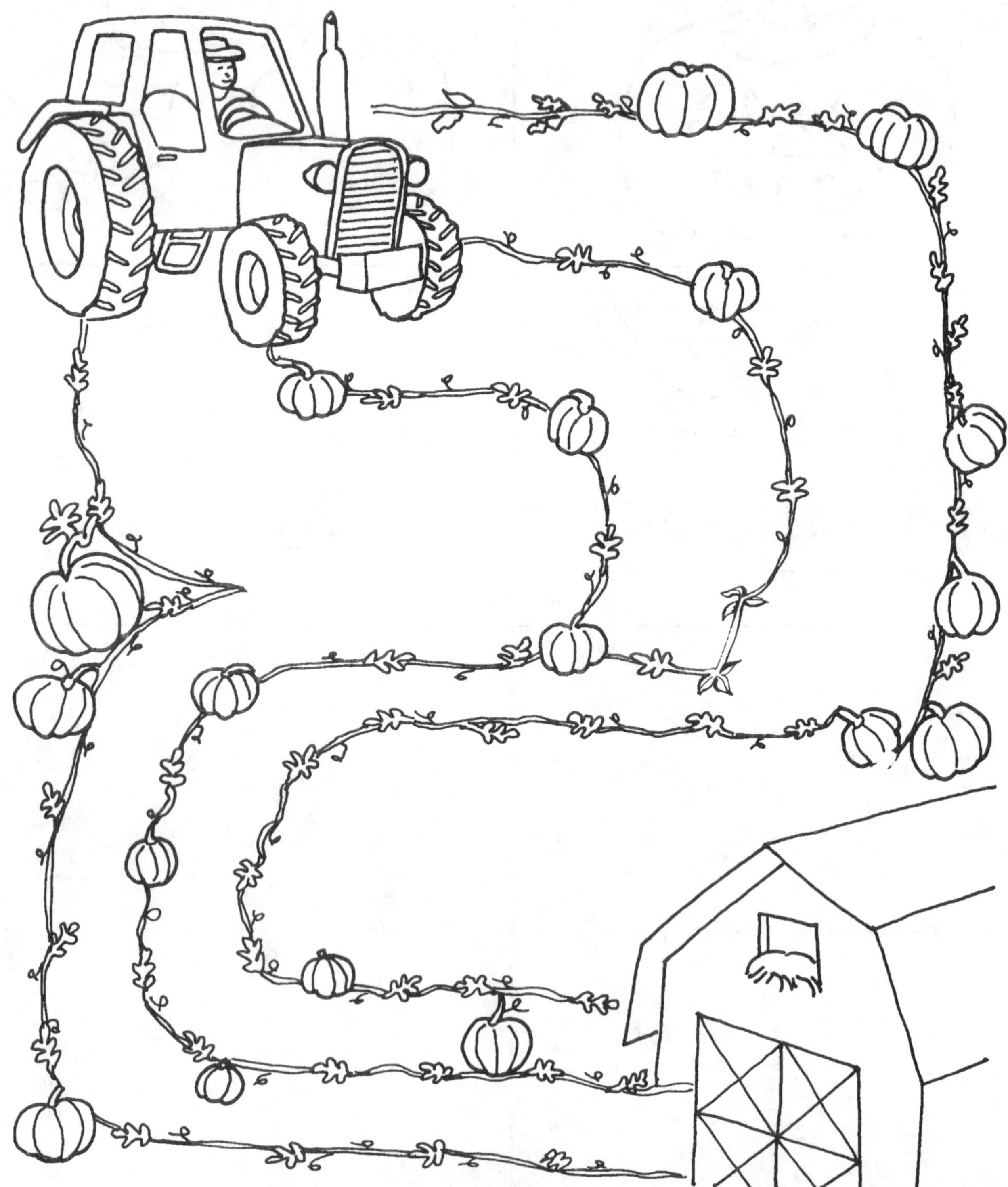

 

Name ______________________________

# Planting Seeds and Root Vegetables

Follow these directions to plant your seeds.

**You will need:**

- potting soil
- watering can
- seeds from several different fruits–oranges, apples, or lemons
- one small pot or a clear plastic cup for each kind of seed

**Directions:**

Dry the seeds overnight on paper towels. Fill each pot or plastic cup with soil.

Use a pencil or your finger to make a hole about two inches (5cm) deep in the soil. Put one seed in each hole and cover it with dirt. Put the pots or plastic cups in a warm, sunny window. Keep the soil moist. Draw three pictures to show what you did.

| First |
| --- |
| Next |
| Last |

# Planting Root Vegetables

Follow these directions to plant your root vegetables.

**You will need:**

- a shallow dish
- sharp knife (teacher only)
- water
- the tops of several root vegetables such as, carrots, beets, or turnips

**Directions:**

Trim all the leaves from the vegetables. Cut about one inch (2.5 cm) off the top of each vegetable. Add enough water to cover the bottom of the plate. Stand the vegetables on end in the plate and place in a sunny window. Be sure there is always water in the plate. Draw how you think your plants will look after weeks one, two, and three.

| Week 1 | Week 2 | Week 3 |
|---|---|---|
| | | |

Check how accurate your predictions were.

**Suggestion:** Use poster size paper and divide it into thirds. Have the children help you illustrate their predictions. You may also have the children draw their predictions on paper.

 

# *My Farm*

## *by Alison Lester*

### Summary

*The author recounts events from her life while growing up on a cattle farm in Australia. Each season the family had different responsibilities for their animals. Their chores, though difficult and tiring, were undertaken with a good-natured spirit.*

### Concepts:

- A child can help with farm chores.
- Cattle need special care.
- Changes in weather affect farm life.

### Concept Activities:

- Discuss with children the chores that the children do on the farm. Ask what chores your students do at their homes. How do they feel about helping?
- If you lived on a farm, what do you think you would like to do?
- Role play a variety of farm chores for your friends to guess.
- Compare life in the country and city. Where do your students live? What kinds of things do they see or do each day that are the same or different from what a country or city child sees or does?
- Recall instances of special care given to the cattle giving birth or to an orphaned calf. Explain how the author was able to help.
- Discuss how weather affected the movement of cattle and the work that needed to be done. Make a chart of farm chores in summer, autumn, winter, and spring.

### Additional Activities:

1. Which animal is your favorite? Why? Use the writing paper on page 21 to write about your favorite animal. Another day, use the writing paper for a story about visiting or living on a farm.
2. Provide watercolors and paper for students to make paintings of animals similar to those in the book. Display them with the original writing assignments above.
3. Brainstorm a list of foods that are made from apples. Make a list of other fruits that grow on trees.
4. Explain the special Australian terms listed at the back of the book. Make a wall chart to show the vocabulary words and pictures of Australian and American words.
5. Glue together the tractor parts from page 27 onto the tractor shape on page 28. Share *Tractors* by Graham Thompson, Gareth Stevens, 1986, to determine the importance of farm machinery.
6. Complete the barn dot-to-dot picture on page 25.

# My Farm *(cont.)*

## Additional Activities *(cont.)*

7. Read more about wombats. What will the family probably do with the baby after it is grown? Why is it wise/unwise to adopt wild animals? If you wish, introduce children to the unusual animals of Australia.
8. What do you think the author and her family did in their "pioneer settlement"? Study the illustration and make a chart or Venn diagram to compare their experience to an American family camping trip.
9. Complete the story about the farm by adding initial consonants on page 22. Write another story about your day. Try to sequence at least five things that you do each day.
10. Talk with the students about their cousins. What do they enjoy doing together? Compare and contrast what the cousins see and do when they visit the farm at Christmas using a Venn diagram or through discussion.
11. Use the work sheet on page 33 to show the foods the mother might send for an afternoon tea or dinner by the seaside.
12. Complete the work sheet on page 26, matching animal parts.
13. Make the *My Farm* booklet, page 29. Complete the illustrations to match the boldfaced words.
14. Prepare the cow counting activity on page 23 for practice in counting sets. If you wish, add number words to the fence sections for reading practice. Share the book *Counting Cows* by Woody Jackson, HBJ, 1995.
15. Ask children to recall the dog in the story. What is her name? (Sadie) What does she do? How many students have a dog? What might be different about the life of a country and a city dog? What do you think will happen to Sadie's puppies as they grow? Complete the work sheet on page 34.
16. Do you think this was a close family? Would you have wanted to live as they did? Explain. Discuss and compare the things that were important to them and to your students.
17. Use the work sheet on page 32 to practice reading and writing numbers.

## Culminating Activities:

1. If possible, arrange for your students to bring packed lunches and enjoy a picnic in a nearby park. Take a walk and look for wildflowers, insects, and small animals. Perhaps you will want to feed the birds. Be sure your students look at the colors of the sky and natural surroundings.
2. Paint a mural of an outdoor farm scene similar to the illustrations in the book. Be sure to fill the whole paper with color. Attach vocabulary words written on index cards to the mural. Children enjoy reading the names of things they have learned about.

 

# My Favorite Farm Animal

**To the teacher:** After the writing page is completed, have children cut the same shape out of red construction paper. (You may also trace the shape onto a ditto master to run directly onto the construction paper.) This will become a cover for the writing page. Cut out a barn door in the front paper to open and reveal the animal picture. Staple along the left roof line.

Name ______________________________

# Initial Consonants

Fill in the correct letters. Use the picture word bank below to complete the words in the story.

## MY FARM

I live on a farm. There are ____ows, ____eep, ____ens, and many ____orses. We all have jobs to do. ____other milks the cow. I feed the ____ickens. The animals live in a ____arn. It is fun to live on a farm.

## Picture Word Bank

sheep

barn

Mother

chickens

hens

cows

horses

# Counting Cows

**To the teacher:** Make this activity for counting practice.

## You will need:

- ten cow patterns from the following page
- 40 spring clothespins
- a black marker or crayon
- 10 squares of white tagboard, 6" x 6" (15 cm x 15 cm)
- masking tape

## Directions:

Copy the cows onto heavy paper. Cut out and spot the bodies to represent sets 1–10. Attach the clothespin legs and stand the cows on the table. Draw a section of farm fence on each tagboard square. Print the numerals 1–10 on the fence sections. Tape the fence together in order, accordion fold it, and stand it on a table. Students match the cow spots to the correct numerals on the fence.

# Counting Cows *(cont.)*

Patterns

Name ______________________________

# Where Do the Animals Live?

Connect the dots and then color the picture.

The animals on the farm live in a ____________________.

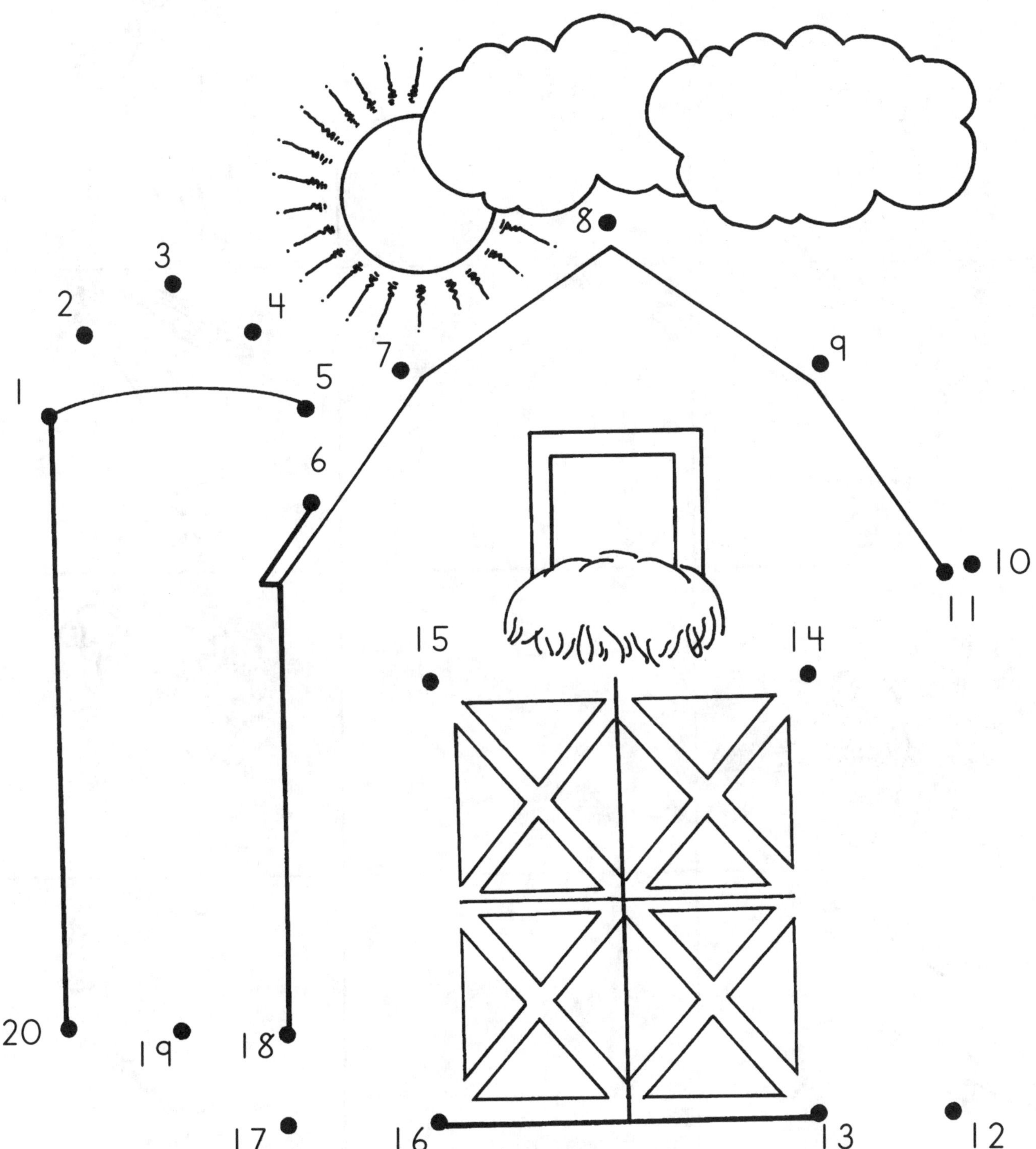

Name ______________________________

# Match the Animal Parts

Color, cut, and glue the back half of the animal to the matching front half.

Can you name the animals out loud?

Name 

# Tractor Activity

1. Color the shapes.
2. Cut out the pieces and match the shapes on the following page.
3. Point to each shape and practice saying its name.

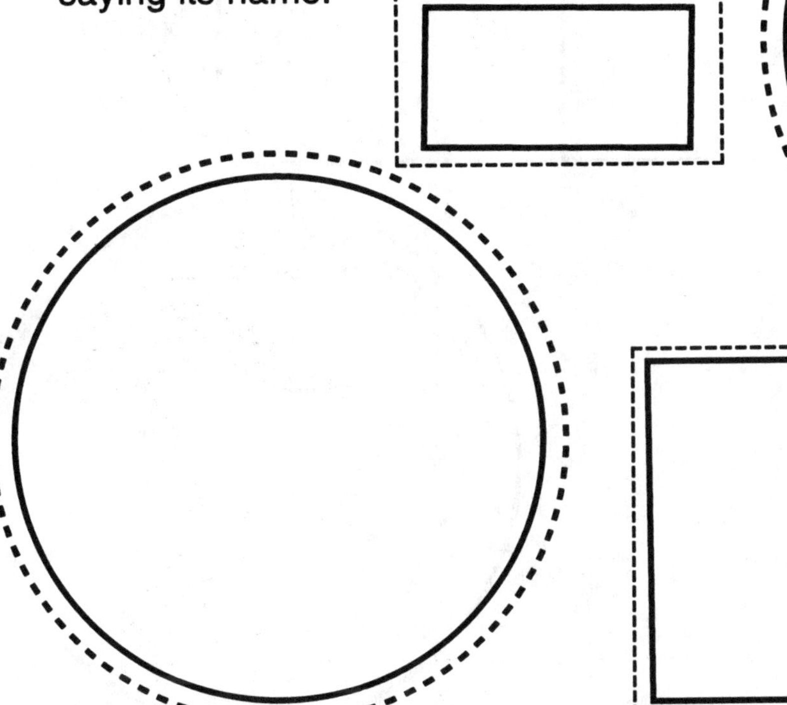

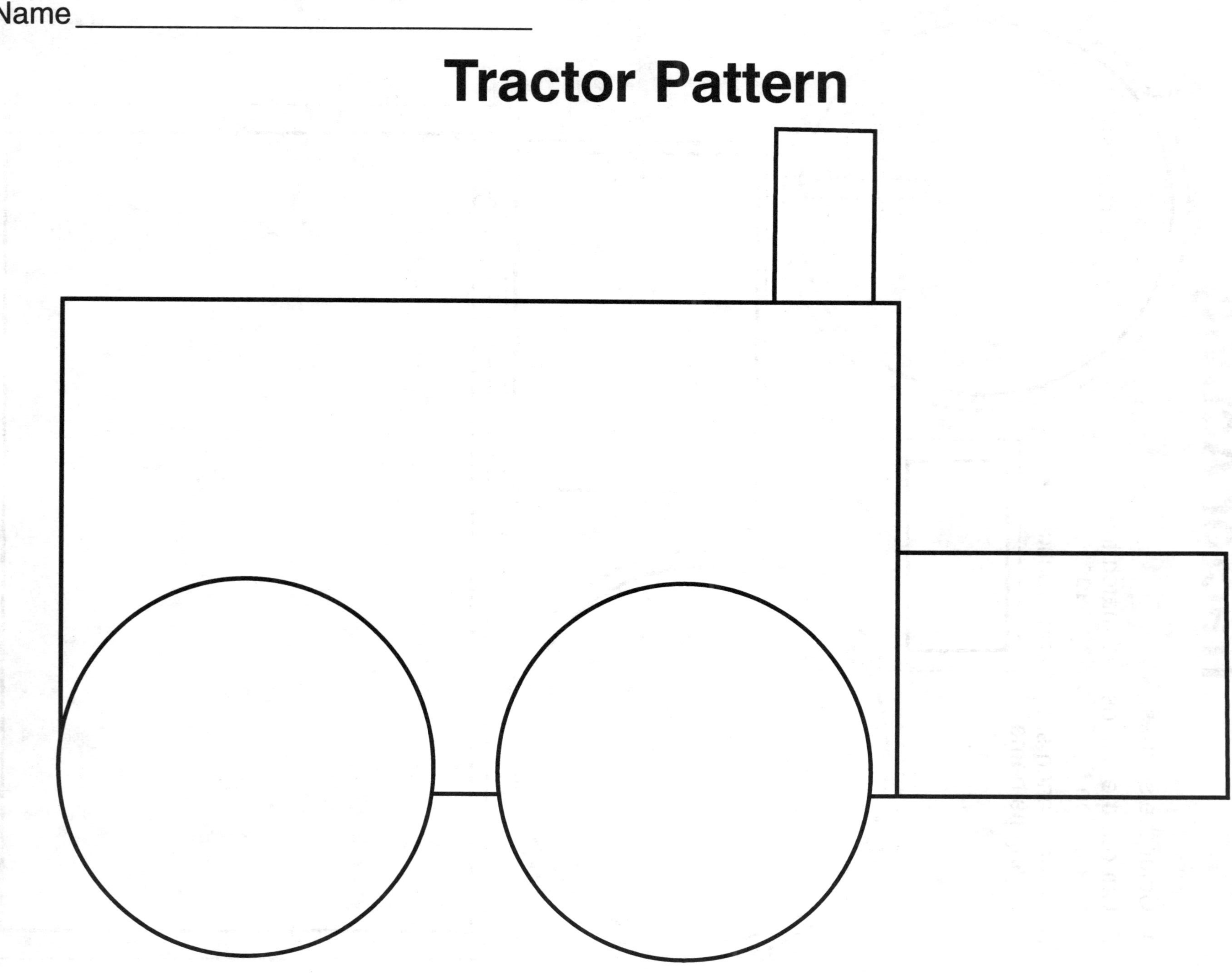
Name ____________
Tractor Pattern

# My Farm Booklet

**Directions:** Reproduce the pages for each child in the class. Let children color the pages and add illustrations for the boldfaced words. Staple them into a book. Give each child a piece of paper cut to size to use to design a cover for his or her book. The children may practice reading and tracking with their completed booklets. This may also be used as a choral reading activity.

# My Farm Booklet *(cont.)*

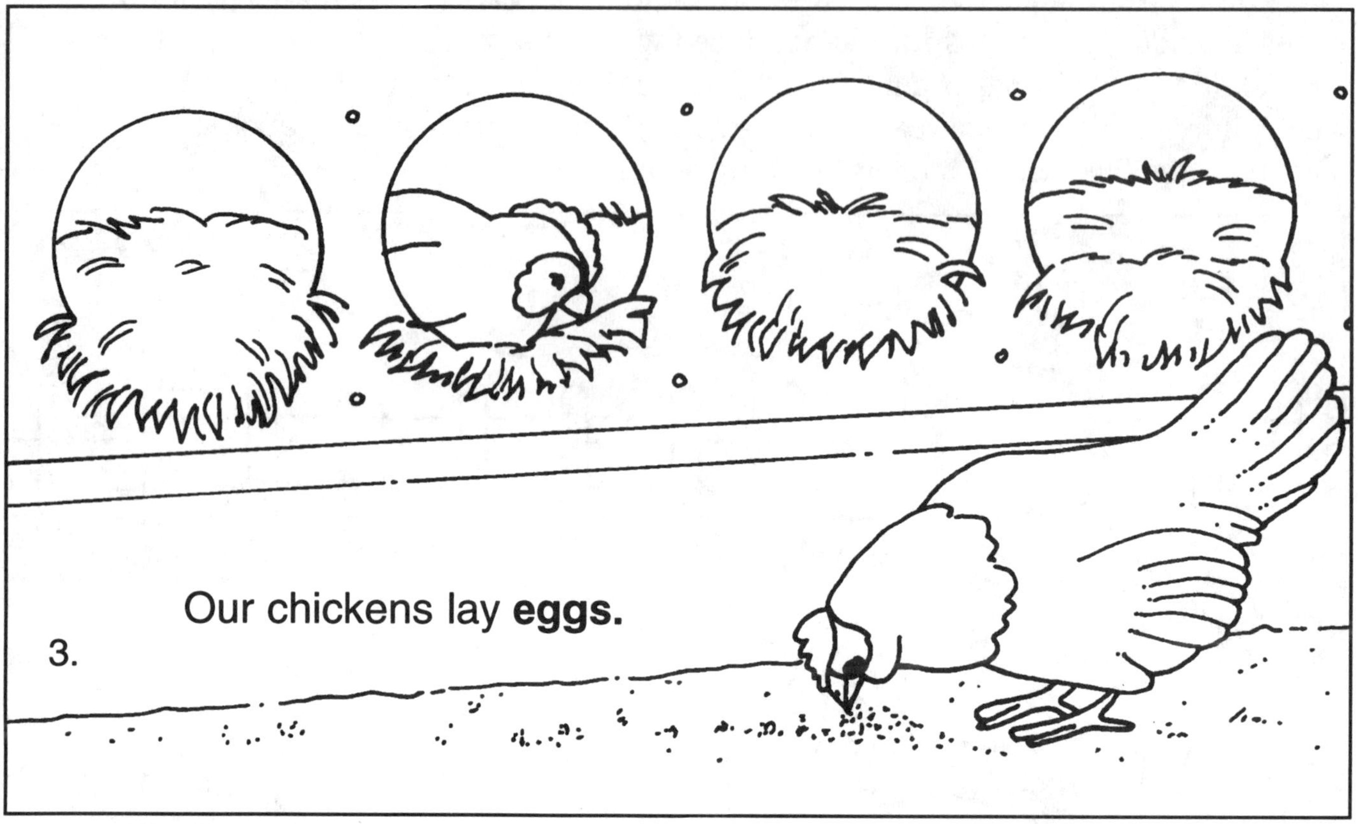

3\. Our chickens lay **eggs.**

4\. We pick **apples** from the orchard.

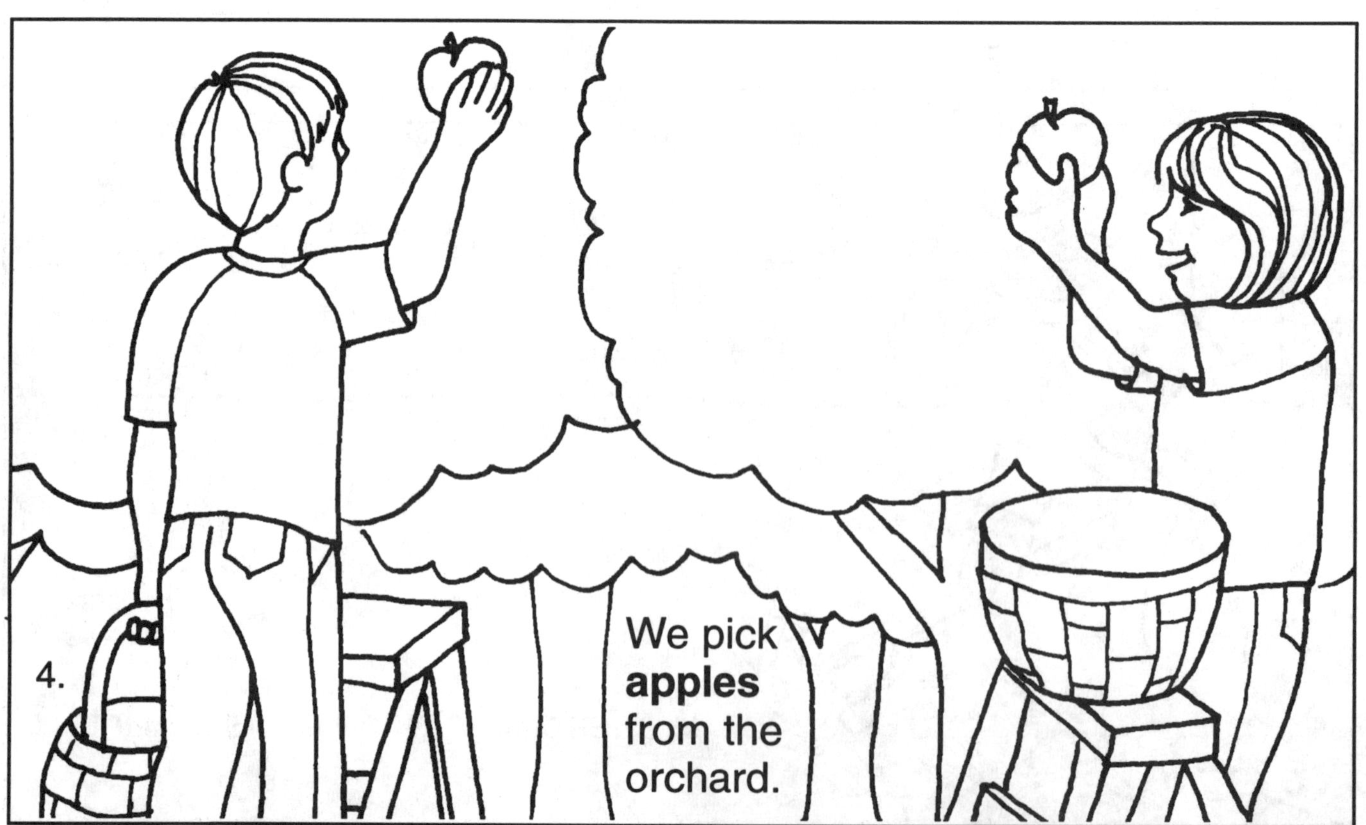

# My Farm Booklet *(cont.)*

Name________________________

# Eggs and Apples

Read the numbers and draw the eggs to represent that number.

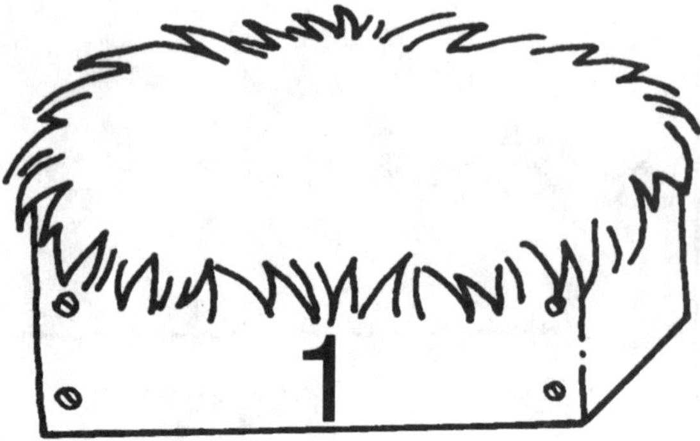

Count the apples and print the numbers on each basket.

Name ____________________________

# Afternoon Tea or a Seaside Dinner

Draw four foods that Mother might have packed for afternoon tea or dinner by the seaside. You may also find foods from a magazine and cut them out to place into the basket. Cut and glue part B on top of part A. Lift to see what's in the basket.

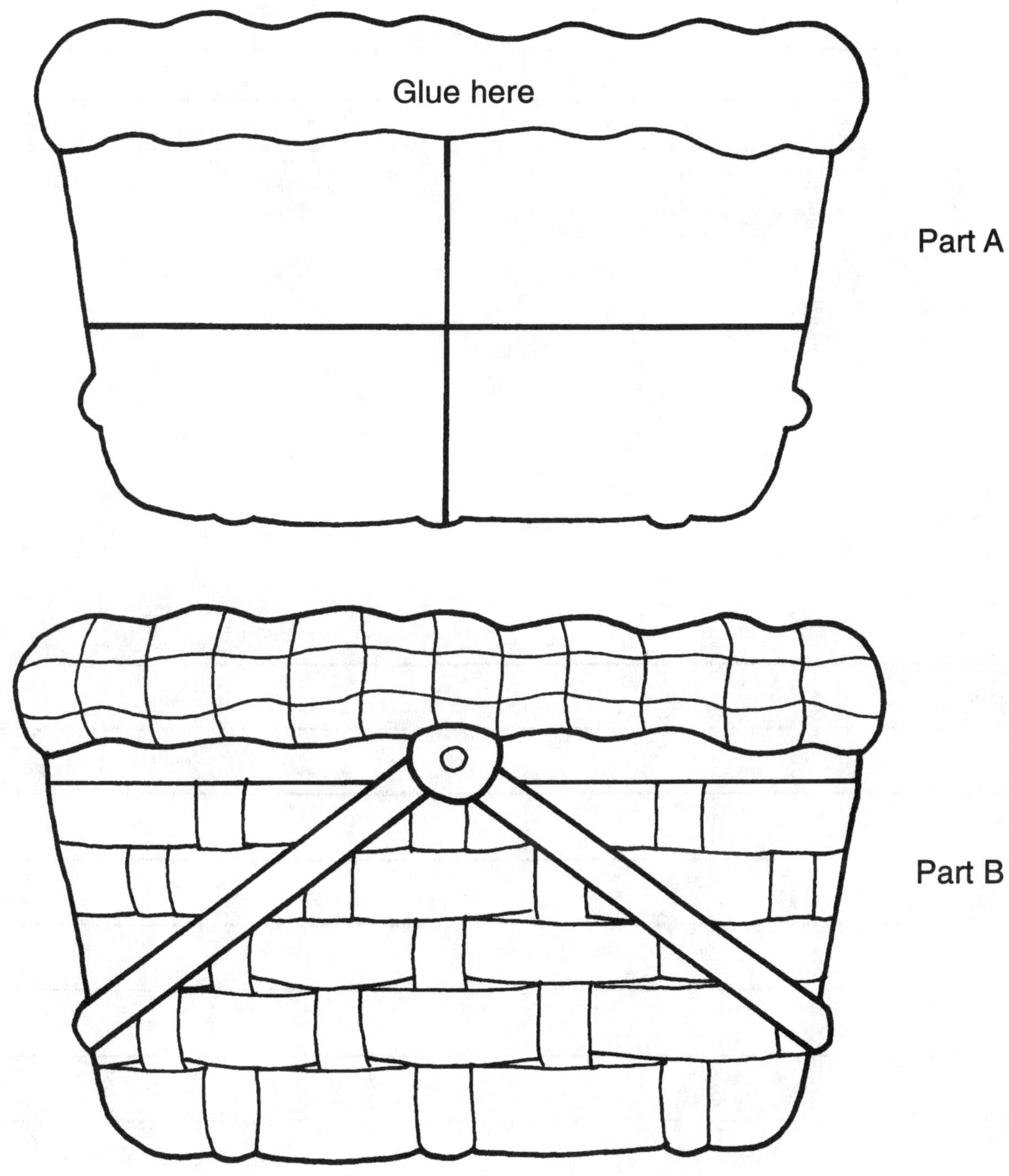

Name ______________________________

# Sadie's Puppies

Sadie had six puppies. What would be good names for them?

**Directions:** Choose one puppy to write about. Illustrate something it might do on a farm.

**Names for puppies**

1. ________________________
2. ________________________
3. ________________________
4. ________________________
5. ________________________
6. ________________________

 

# *The Year at Maple Hill Farm*

*by Alice and Martin Provensen*


## Summary

*In this classic book, children will see the effects of seasonal weather changes on farm animals, birds, and insects. Specific farm chores are shown for each month.*


## Concepts

- Weather changes affect animals living on the farm.
- The twelve months are divided into four seasons.
- Animals must be cared for year round.

## Concept Activities:

- Make children aware of the weather through discussion each day. Chart the weather for a given period of time.
- How do weather conditions change the way they feel? dress? How does weather control what they do? What do they think farm animals do in warm weather? cold weather? Complete the work sheet Seasons All Around Us on page 40.
- Discuss how animals adapt to weather changes. If possible, show pictures.
- Look at a calendar to find the current month. Name as many months as possible. Determine months with important dates: When is your birthday? When is Thanksgiving? When is Halloween?
- Describe the weather for each season. Allow students to vote for their favorite season, because of weather or special days. Graph the results.
- Compare/Contrast two seasons on a chart by using a combination of pictures and words.
- If children have pets, ask them what they do to care for them. How are those responsibilities similar or different from caring for farm animals? Use a chart/Venn diagram to compare and contrast.

## Additional Activities:

1. Use the vocabulary cards on page 37 to practice the names of the months. Make available calendars showing seasonal pictures. As a class, practice saying the months in order. This may be done daily as part of your morning opening.
2. Share the book *Chicken Soup and Rice* by Maurice Sendak, HarperCollins Publisher, 1962. This is a small book of poems for each month. Use the current poem each month for printing practice.
3. Ask each student to make an illustration of his or her favorite month on 9" x 12" (23 cm x 30 cm) drawing paper. Use vocabulary cards from page 37 and glue them on the name label. Organize the pictures in order and glue onto large paper to make a display similar to the book cover.
4. Complete My Book of Seasons on page 46. Ask students to draw themselves doing some activity in each illustration. Later, have an adult record their stories in the books.

# *The Year at Maple Hill Farm* *(cont.)*

## Additional Activities *(cont.)*

5. Read the F-A-R-M poetry on page 38. Explain that this is a poetry style that does not need to rhyme. Write several additional examples as a group and use one for printing practice, using the work sheet on page 39.
6. If it is spring, discuss the birth of plants and animals taking place in nature. Match the animal babies to their parents on the work sheet on page 43.
7. Practice counting sets with the leaves and flowers on page 44.
8. Complete the work sheet sequencing a baby chick from an egg to a hen on page 45. Encourage children to make similar charts for other animals.
9. Brainstorm a list of animals from the book. Show the names of both adult and baby animals. If possible, include picture cues so that your students can "read" the list. Compare a baby animal to an adult animal. Do the activity on page 49. Discuss how the sheep looks before and after it grows wool.
10. Share the poems, "Four Seasons" (Anonymous) and "The Months" by Sara Coleridge, *Random House Book of Poetry for Children*, 1983. Write your own rhyming couplet for each month and combine them into a poem similar to "The Months."

## Culminating Activity:

1. Prepare and play the Animal Lotto game, page 52.
2. Make an animal puppet with a lunch-size paper bag; use the patterns on pages 54 and 55. Ask students to say riddles or write poems about their puppet. Younger children may use the puppets and make animal sounds. Sing "The Farmyard Song" (See Resources) and share the book *A Farmyard Song* by Christopher Manson, North-South Books, 1992.
3. Use the patterns on page 51 to build a farm. Put all the pieces together in a floor display. Count and categorize the animals and plants as many ways as possible. Small index cards make good labels. Have children transfer the "farm" onto paper.

# Vocabulary Cards

| | |
|---|---|
| January | July |
| February | August |
| March | September |
| April | October |
| May | November |
| June | December |

# Seasons on the F-A-R-M

## Preparation:

- Read the sample poems and point out the **F-A-R-M** structure.
- Explain that these poems do not need to have rhyming words.
- Discuss what each poem tells about the seasons on the farm.
- As a class, write your own set of poems for farm seasons. Individual students can use the form on page 39 to write F-A-R-M poems of their own.
- Extend this activity by having students write poems about themselves, based on the letters of their first names.

**Spring**

**F**ields are plowed,
**A**nd seeds are planted. Soon
**R**ains will come
**M**aking everything green.

**Summer**

**F**ood is growing and
**A**nimals are grazing, getting
**R**eady for
**M**arket.

**Autumn**

**F**ields of wheat,
**A**utumn's harvest,
**R**estless birds prepare to
**M**igrate south.

**Winter**

**F**rozen ground,
**A**nimals are inside,
**R**esting until
**M**arch brings sunshine.

 

Name ______________________________

# Seasons on the F-A-R-M *(cont.)*

Create a poem of your own and color the barn.

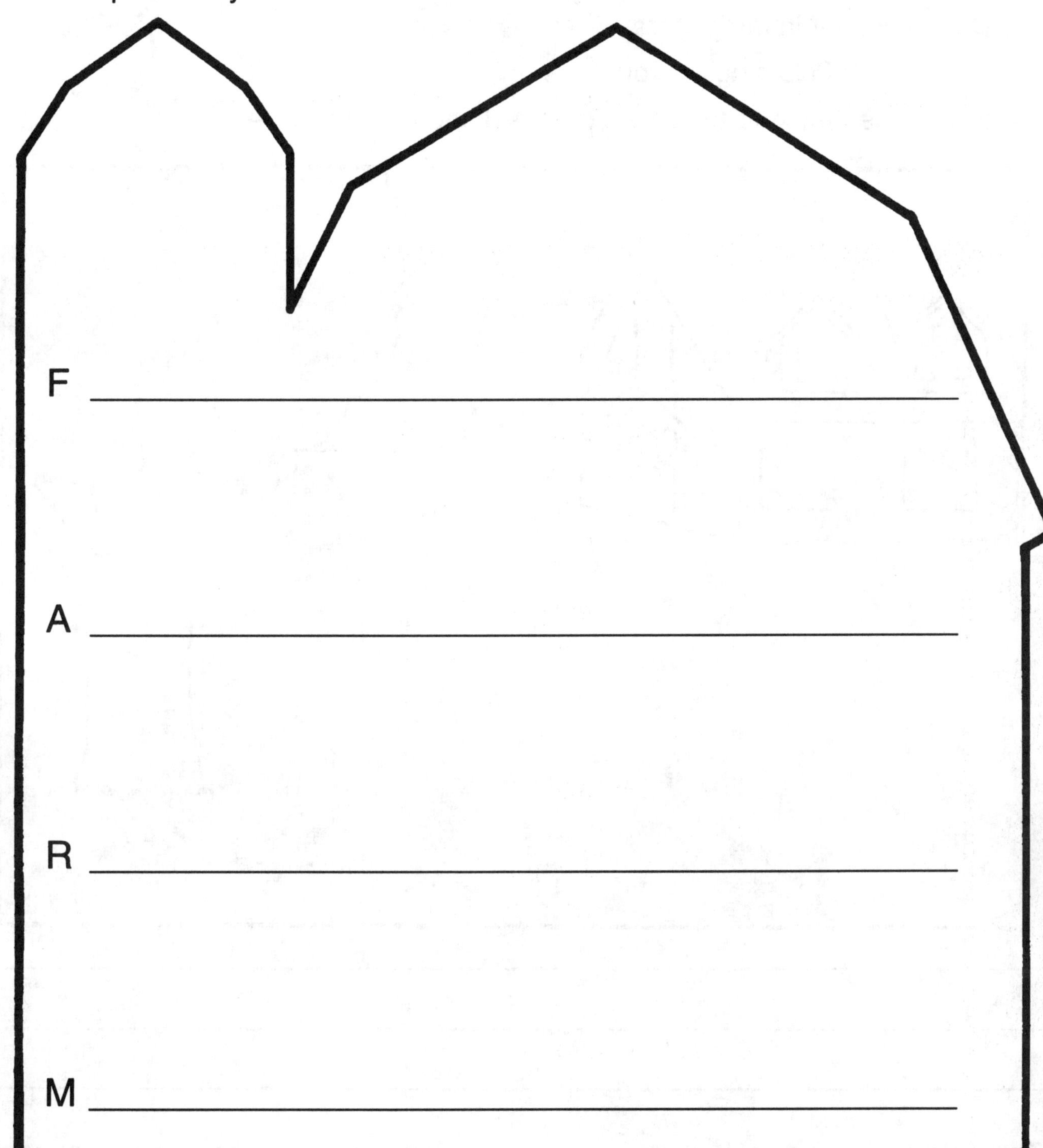

Name ______________________________

# Seasons All Around Us

1. Color the picture to show what the weather is like now where you live.
2. Draw yourself in the picture.
3. What type of clothing do you have on?
4. Write a sentence to tell what you are doing in the picture.

_______________________________________________

-----------------------------------------------

_______________________________________________

**Extension:** On the back, draw a picture of a farm, showing the current season. What is the farmer doing? What are the animals doing?

 

Name ______________________________

# Seasons of the Year

Reproduce the page for each child in the class. Let children color the pictures below, cut them apart, and glue them by the correct season.

Name ______________________________

# Can You Help the Farmer?

Reproduce the page for each child in the class. Let the children color the pictures below, cut them apart, and glue them in the correct place.

| Inside | Outside |
| --- | --- |

Name ________________________________

# Animal Babies

Draw a line to match the baby animals to the adult animals. Do the animals look similar to each other?

**baby animals** | **adult animals**

Name ______________________________

# Counting Leaves and Flowers

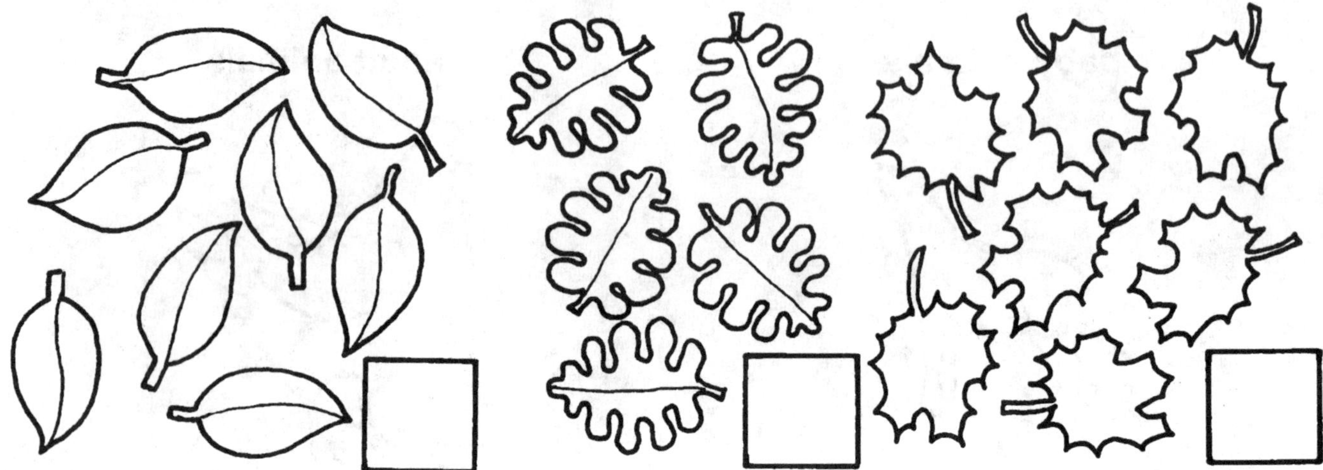

Count the leaves and write the numbers.

---

Read the numbers and draw that many flowers.

| 6 | 4 | 9 |
|---|---|---|

 

Name ______________________________

# Which Came First?

Number the events in their correct order. Color each picture.

# My Book of Seasons

**Directions:** Reproduce the pages of the booklet for each child in the class. Let children color the pages. Cut apart and staple them to make a book. Give them a sheet of paper cut to size and have them design a cover. Have children practice reading the seasons of the year as well as associating the names with the pictures.

Spring

Winter

 

# My Book of Seasons *(cont.)*

Summer

Autumn

Name ______________________________

# The Good Old Brown Hen

1. Tear scraps of brown paper for feathers.
2. Glue feathers on the hen.
3. Color the baby chicks.
4. Retell the part in the story about **The Good Old Brown Hen** to a friend. Describe what the hen looks like after completing the above activity.

# Before/After

**You will need:**

- two copies of the sheep
- scissors
- cotton balls
- crayons
- glue

1. Color the sheep on both pages.
2. Cut out the sheep pattern on this page.
3. Glue it onto the sheep on the next page.
4. Glue the cotton balls on the sheep on top. (That is your **before** picture.)
5. Lift the top sheep up and notice what happens to the sheep **after** its "haircut."

# Before/After *(cont.)*

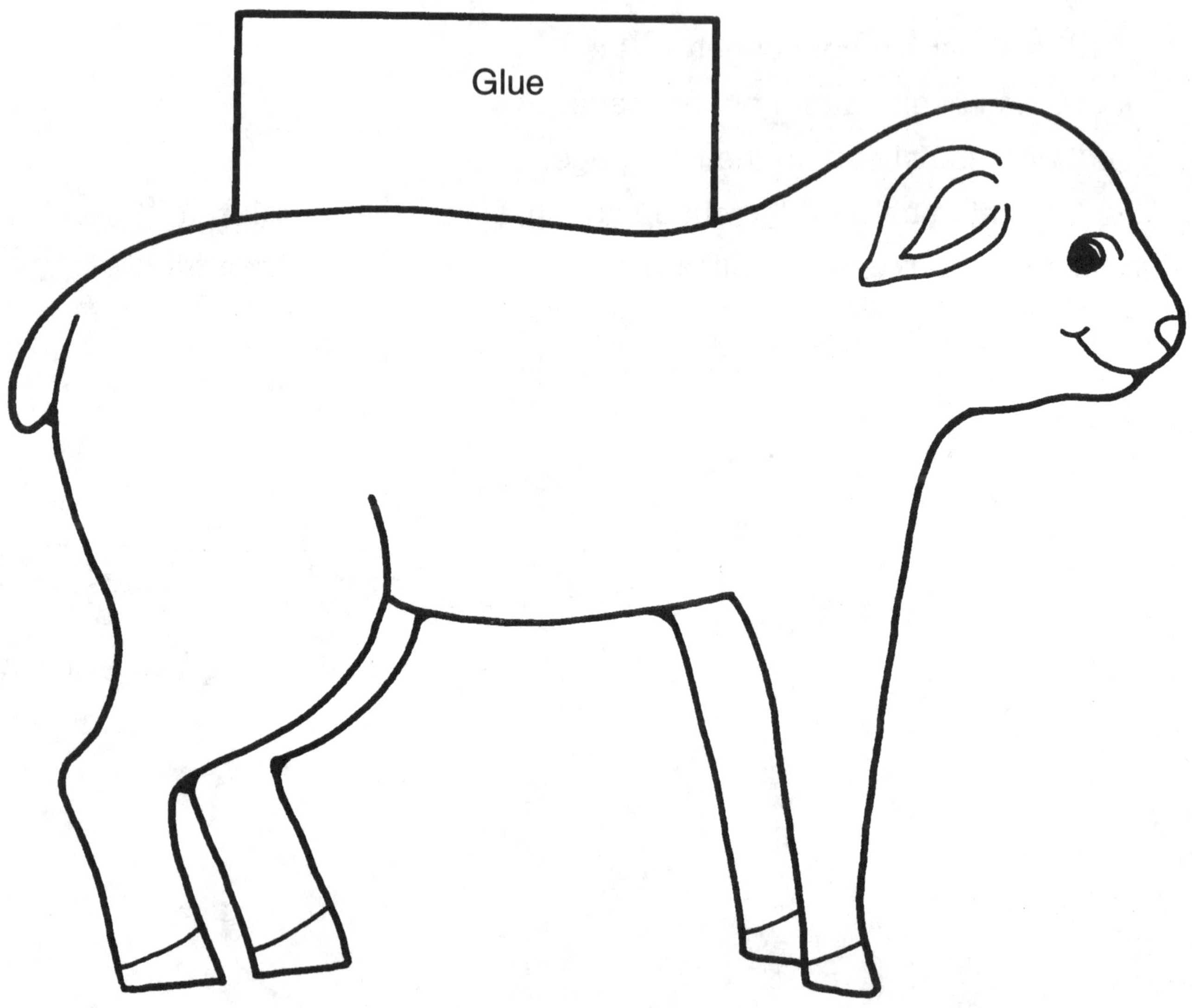


When does the sheep get a "haircut?"

 

# Build a Farm

**You will need the following:**

- lots of cardboard tubes, cut in 3"x 6" lengths (7.5 cm x 15 cm)
- heavy drawing paper
- stapler
- crayons
- scissors

**Directions:** Your farm may include vegetables, animals, or both.

Children may work in groups to draw assigned pieces. Cut around the pictures and staple them to short lengths of tube so they will be free standing.

**To use:** This will make a large floor display for a free-time activity center. You may add brown paper for fields and blue paper for a pond. Complete your "Maple Hill Farm" with a house and barn made from shoe boxes and a free-standing farmer and tractor. Children may count and categorize the animals and plants in many ways: by color, size, sound, etc. Use the farm for language development and story writing by changing the scenes periodically.

# Animal Lotto

A game for four players.

**You will need the following:**

- four pieces of construction paper
- markers
- glue
- ruler
- tagboard for game cards and markers
- scissors
- five sets of animal pictures
- crayons

**Directions:** Fold a sheet of construction paper into 12 squares.

Copy, color and cut apart the animal pictures on page 53. Glue them in each square of the construction paper. Be sure each card is different. Laminate. Glue a complete set of animal pictures to tagboard, laminate, cut apart, and store in a separate container. If appropriate, you may substitute word cards for the animal pictures. Color and cut extra tagboard into 2" (5 cm) squares to use as game markers.

**How to Play:** Students in turn choose a picture or word card from the container. Everyone covers that animal if it appears on his or her card. The first player to cover four animals in a row is the winner.

# Animal Lotto *(cont.)*

**Animal Picture Cards**

# Patterns for Animal Puppets

Use these patterns to attach to paper lunch sacks to make animal puppets. You may use these puppets for oral language activities.

See page 36.

# Patterns for Animal Puppets *(cont.)*

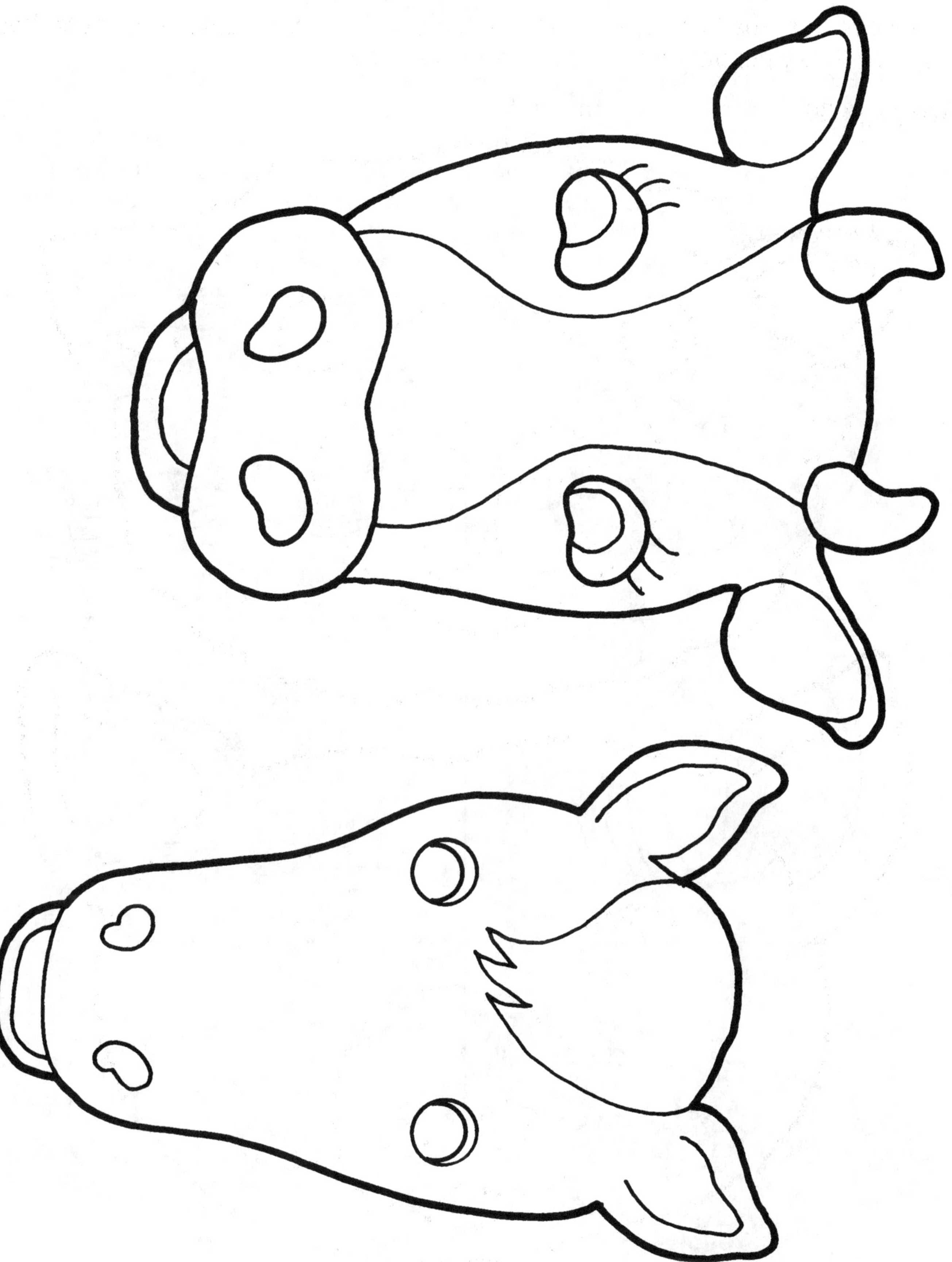

Name ________________________________

# Colorful Food

**To the teacher:** This activity will provide an opportunity for your students to associate color words with common foods. Before giving out this work sheet, use "color" or word cards and play a game with your students. **Example:** Display "color" or word cards for all to see. **Say:** I am thinking of a food from which we cut out a jack-o'-lantern face at Halloween time. What is it? What color is it? After playing the game, distribute the work sheet. When the work sheet is completed, the children may play this game with a partner.

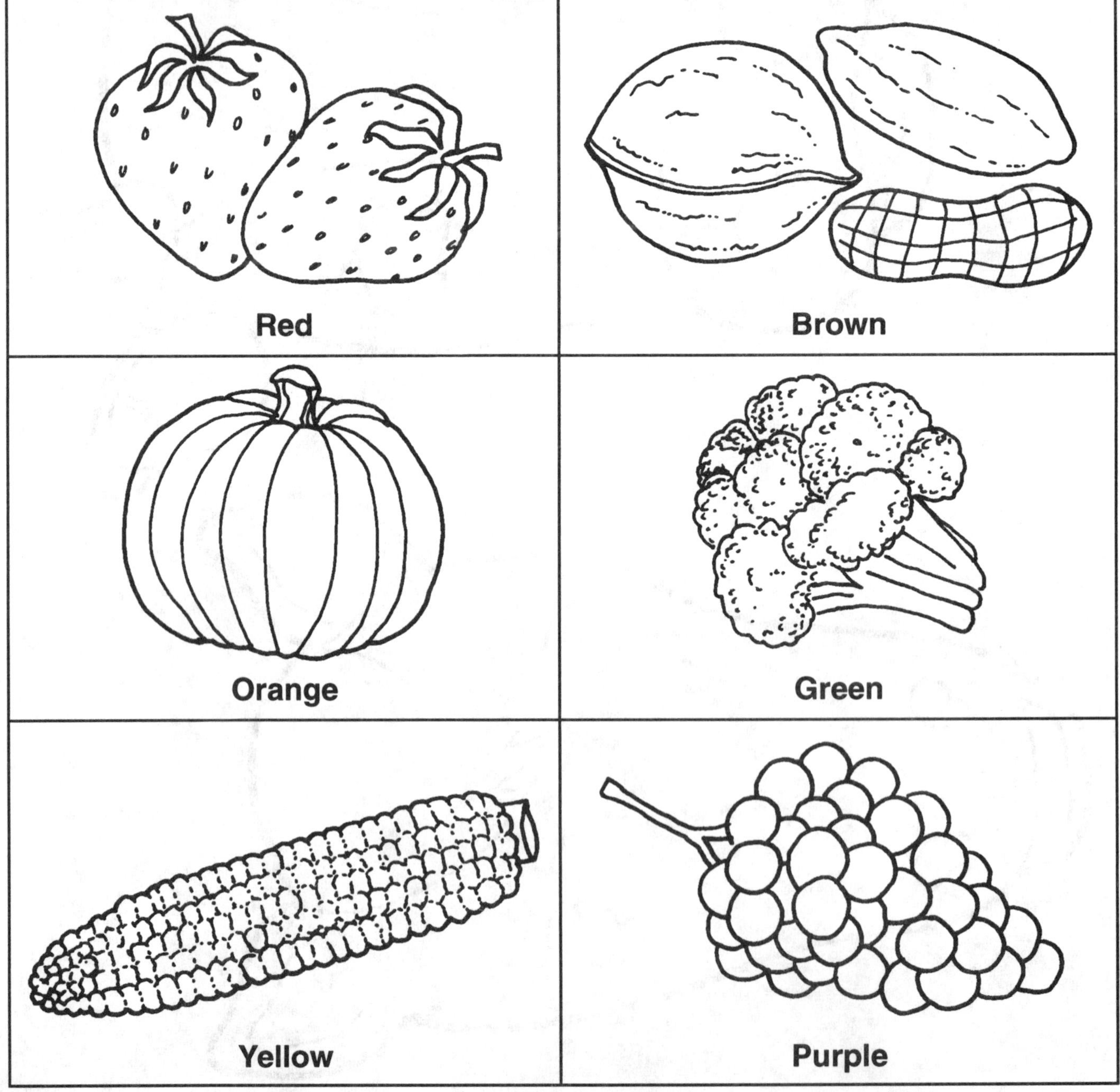


 

# Farm Colors

**Read and do the following.**

1. Draw yellow grain for the hens to eat.
2. Draw black spots on one cow.
3. Draw a brown fence between the cows and the hens.
4. Color the barn red.

How many? ________ ________ ________

Name ______________________________

# What Comes Next?

Draw pictures to illustrate and complete the sentences.

| | |
|---|---|
| The brown cow is eating in the field. | The cow is being milked in the big red barn. |
| The milk from the cow is put in a carton. | I am drinking a cold glass of milk. |

 

Name ______________________________

# What Comes Next? *(cont.)*

Draw a picture to illustrate and complete the sentence.

| | |
|---|---|
| The wooly sheep is in the red barn. | The sheep is being shorn in the red barn. |
| My mother is knitting a sweater for me. | I am wearing my new blue sweater. |

# Can You Find the Opposites?

Circle the correct word and color the picture.

Name ______________________________

# What Letter Is Last?

1. Look at the animal pictures and say each animal name.
2. Listen to the ending sound.
3. Choose the correct letter to complete the animal words.
4. Cross out each animal picture as you complete each word.
5. Use lowercase letters.

| do___ |
|---|
| pi___ |
| duc___ |
| shee___ |
| he___ |
| goa___ |
| rabbi___ |

Name ______________________________

# What Letter Is Last? *(cont.)*

1. Look at the food pictures and say each name.
2. Listen to the ending sound.
3. Choose the correct letter to complete the food words.
4. Cross out each food picture as you complete each word.
5. Use upper case letters.

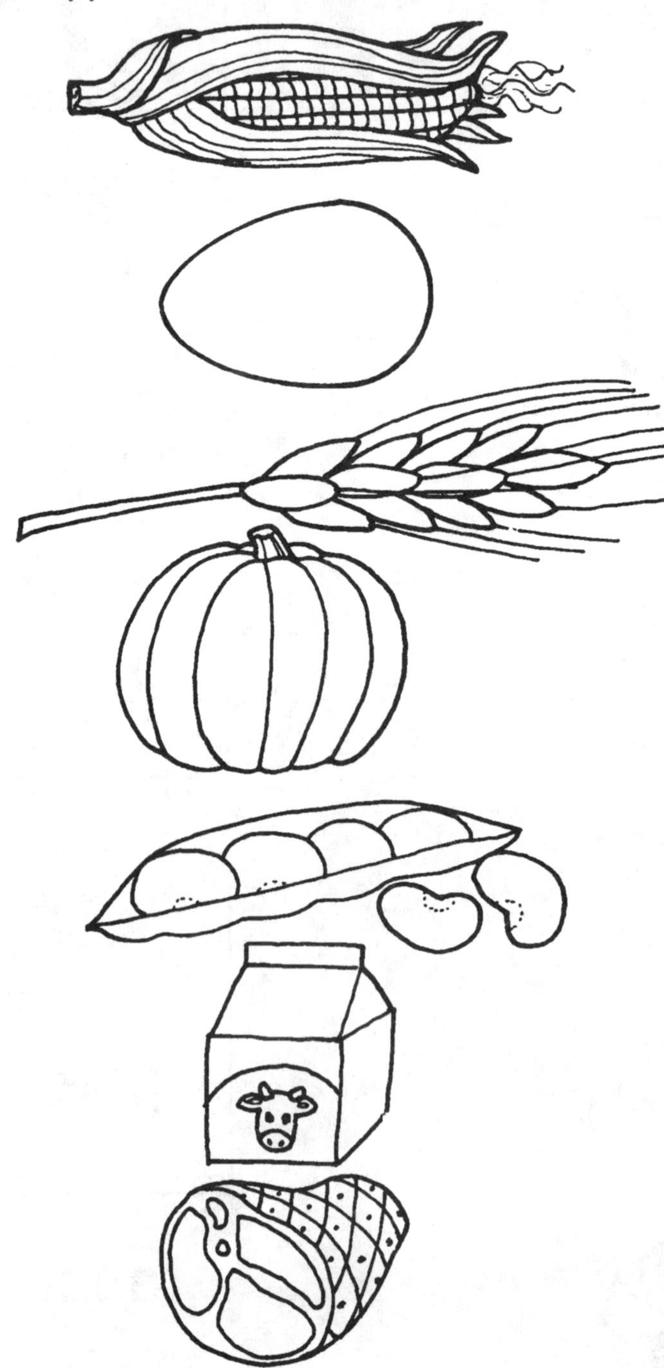

| COR___ |
|---|
| EG___ |
| WHEA___ |
| PUMPKI___ |
| BEAN___ |
| MIL___ |
| HA___ |

 

Name ______________________________

# Hidden Pictures

Find and color 3 orange  s and 3 red  s.

Color the rest of the picture any colors you choose.

*Math/Geometry*

Name ______________________________

# Barnyard Shapes Picture

1. Trace the shapes.
2. Color the picture.
3. Can you find the shapes in the picture below?

Name ______________________________

# Farm Graph

**To the teacher:** Have each student bring in his/her favorite fruit or vegetable that is either brown, orange, red, green or yellow. This may first be done as a group activity, using a floor-size graph. Have the children fill in the graph below with the results.

| | | | | |
|---|---|---|---|---|
| | | | | |
| | | | | |
| | | | | |
| | | | | |
| | | | | |
| | | | | |
| | | | | |
| | | | | |
| | | | | |
| | | | | |
| | | | | |
| | | | | |
| | | | | |
| | | | | |
| brown | green | orange | red | yellow |

Which is more?__________ Which is less?__________

Name ______________________________

# Farm Patterns

Finish the pattern by cutting the pictures from the bottom of the page and matching them to the correct patterns.

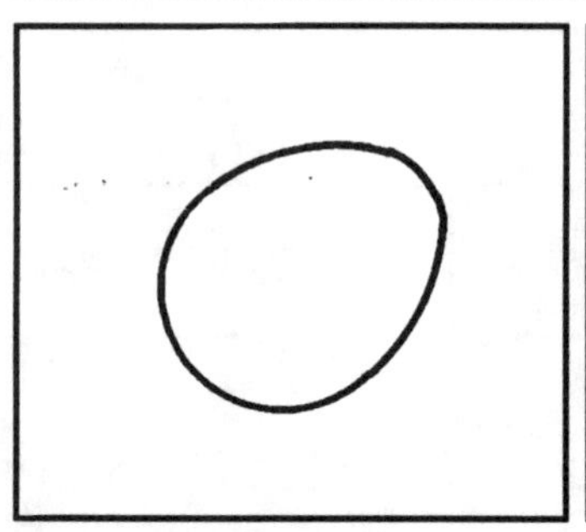

Name ______________________________

# Farm Facts

+ = How many in all? ____________

+ = How many in all? ____________

+ = How many in all? ____________

+ = How many in all? ____________

+ = How many in all? ____________

Illustrate and complete the math problems below.

$$\begin{array}{r} 3 \\ +3 \\ \hline \end{array}$$

$$\begin{array}{r} 2 \\ +3 \\ \hline \end{array}$$

$$\begin{array}{r} 4 \\ +1 \\ \hline \end{array}$$

Name ______________________________

# Food Facts

Add the numbers together by counting the vegetables. Write the answer on the pot.

3 + 4 =

2 + 3 =

4 + 2 =

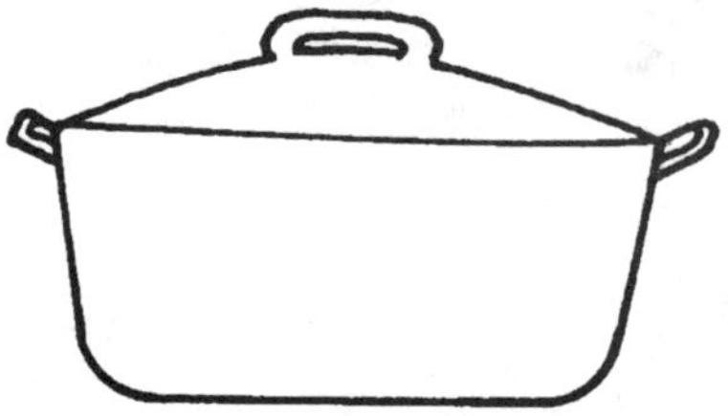

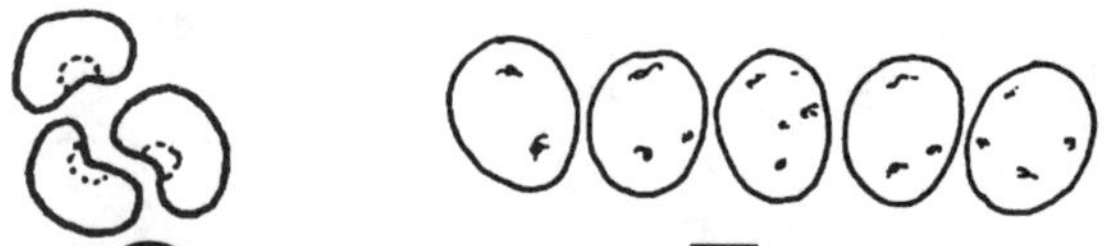

3 + 5 =

2 + 2 =

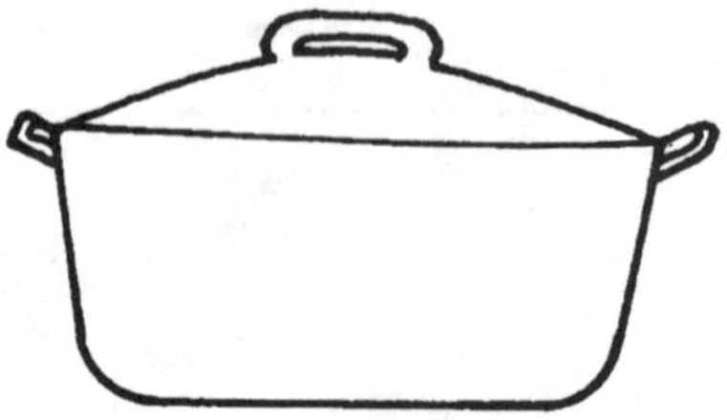

4 + 5 =

Name ______________________________

# Growing Things

Color all the things that will grow and change.

# Hayloft Activity

**Directions:**

1. Color the barn and wagon on page 71.
2. Color and cut out the barn doors on this page. Glue or tape them along the outer edge of the barn on page 71.
3. Color and cut out the hay bales on this page. Glue them in the loft or on the wagon on page 71.
4. This activity may be used as a story starter.

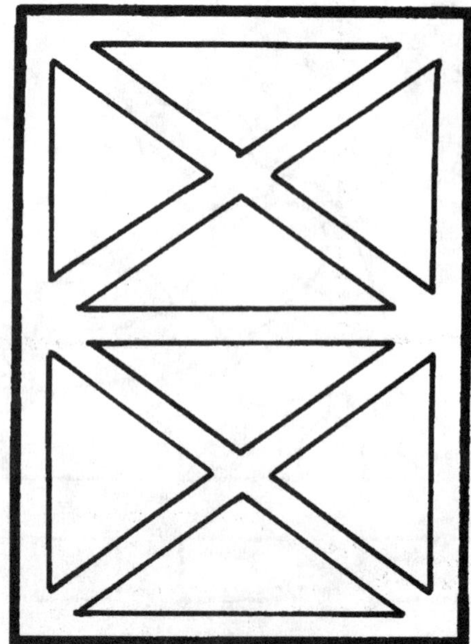

Name ______________________________

# Hayloft Activity *(cont.)*

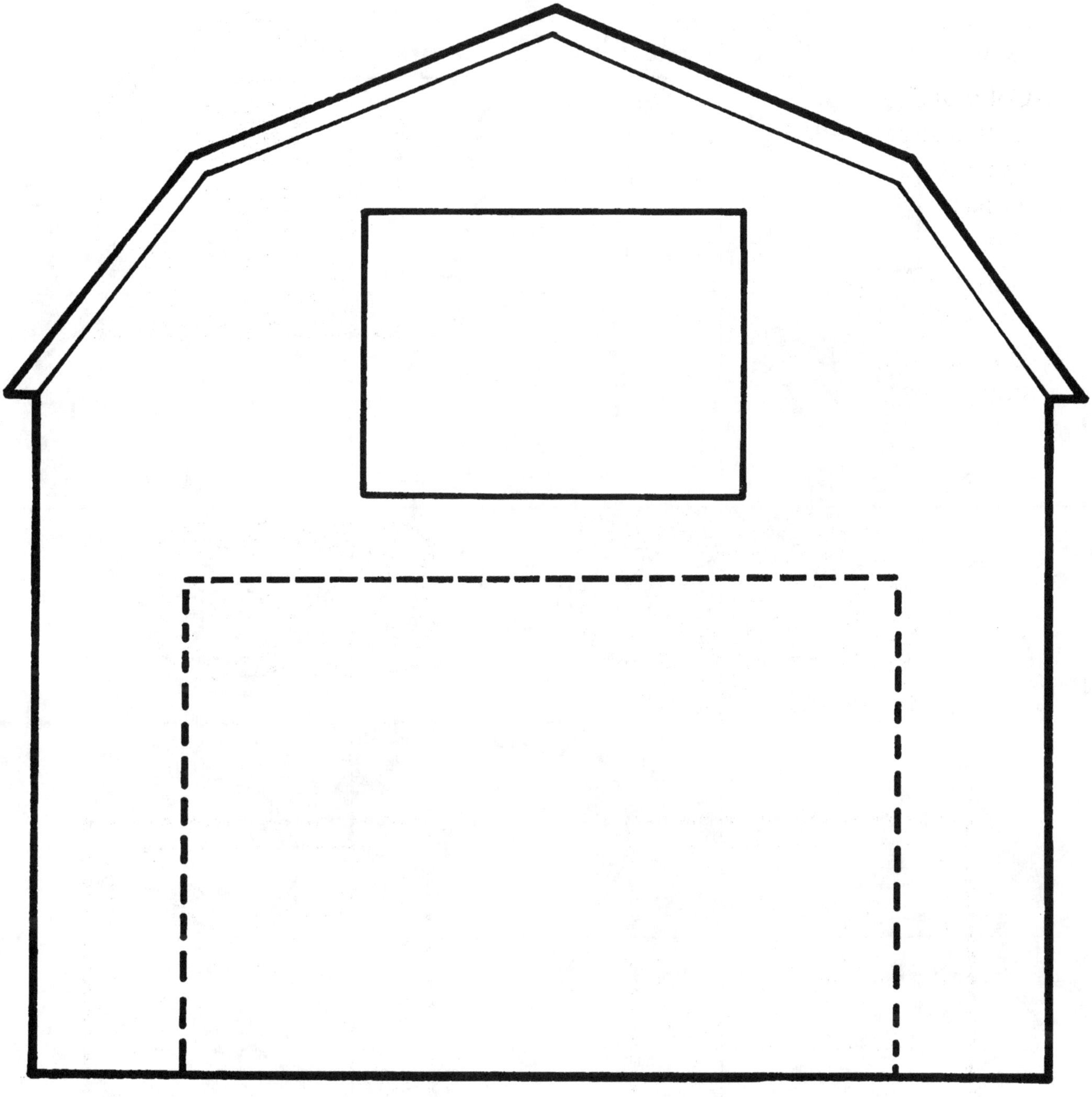

Name ______________________________

# Farm Go-Togethers

Color the two pictures in each line across that go together.

# Animal Names

Print the correct name on each line.

**Word Bank**

**Cow** **Sheep** **Hen** **Pig** **Horse** **Goat**

# Build A Scarecrow

**You will need:**

- two drinking straws
- glue
- scarecrow pattern
- crayons
- small lump of clay
- scissors

**Directions:**

Color the scarecrow, front and back. Cut out both pieces. Glue the straws inside the back piece in the shape of a "T." Glue the front of the scarecrow to the back over the straw. Stand your scarecrow in a lump of clay. This may be used as a centerpiece.

# Build a Scarecrow *(cont.)*

# A Country Breakfast

You may wish to conclude your unit on the farm with a special breakfast. Remind children that farmers eat a large breakfast to give them energy to do their morning work. Here is a recipe that is delicious and easy to prepare.

### Farm Sandwich

- an electric skillet
- a spatula
- a whisk
- a large mixing bowl
- a toaster
- 12 small paper plates and plastic forks

### Ingredients for 12 servings

- 12 eggs
- ½ c. milk (125 mL)
- 2 c. chopped ham (500 mL)
- 2 T. margarine (30 mL)
- 6 English muffins, split and toasted
- ½ c. grated yellow cheese (125 mL)

**Directions:** Saute the ham in melted margarine in the skillet until it is heated through. Beat the eggs and milk together and add to the skillet. Cook over low heat until the eggs are set. Spoon onto the toasted English muffins. Top with grated cheese.

If you prefer, children may help prepare this healthy lunch. Take some time to name the farm products in your soup.

### Vegetable Soup

- a crock pot
- large spoon or ladle
- a can opener
- foam cups and plastic spoons

### Ingredients for 12 servings

- 46 oz.(1.36L) tomato juice
- 2-10 ½ oz. (297g) cans beef bouillon
- 2-16 oz. (454g) bags of frozen vegetables for soup

**Directions:** Put all the ingredients in the crock pot. Turn to high setting. Stir. You may wish to add a soup can or two of water. Reduce to low setting when the soup is hot.

Serve with biscuits or bread and butter. Make your own butter by shaking a small container of whipping cream in a tightly sealed jar for about 20 minutes.

 

# Bulletin Board

These interactive bulletin boards provide decoration as well as a learning center. They are best adapted for a small area and used with a group of 4-6 children.

**You will need the following:**

- white background paper
- scissors
- glue
- crayons
- paint
- blank word cards
- multiple copies of the vocabulary word and picture cards on pages 8 and 9.

**Directions:** Cover the entire bulletin board with white paper. Color or paint a simple background. Color and cut out the barn. Glue it on the background. Add animals, a tractor, and trees according to specific skills. Make word cards for any not provided.

**Suggested Uses:**

1. Attach a season word card to the top of the board. Ask students to create an appropriate scene.
2. Ask one student to display several animals in the farmyard and have another student attach appropriate name cards.
3. Have students demonstrate an understanding of antonyms by placing animals appropriately when shown word cards.
4. Draw or add magazine pictures of children dressed for each season. Display a season word card that matches the children's dress. An example would be a child in a warm coat that matches the children's coat with the word card coat.
5. Arrange an interesting farm scene. Tell a story expressing the action or describing the scene. Print the story on chart paper. Practice reading the story together.

# Songs

***Oats, Peas, Beans***

Oats, peas, beans, and barley grow,
Oats, peas, beans, and barley grow,
Do you or I or anyone know
How oats, peas, beans, and barley grow?
First the farmer sows his seeds,
then he stands and takes his ease,
Stamps his feet, and claps his hands,
And turns around to view the land.
Waiting for a partner,
Waiting for a partner,
Open the ring and take one in,
And then we'll dance and gaily sing.

***Bought Me a Cat***

Bought me a cat, the cat pleased me,
Fed my cat under yonder tree,
Cat went fiddle-i-fee, fiddle-i-fee.
Bought me a hen,
the hen pleased me,
Fed my hen under yonder tree,
Hen went chipsy chopsy,
Cat went fiddle-i-fee, fiddle-i-fee.

***The Farmer in the Dell***

The farmer in the dell,
The farmer in the dell,
Hi ho the derry o,
The farmer in the dell
The farmer takes a wife...
The wife takes the child...
The child takes the nurse...
The nurse takes the dog...
The dog takes the cat...
The cat takes the rat...
The rat takes the cheese...
The cheese stands alone...

***Old McDonald Had a Farm***

Old McDonald had a farm,
e-i-e-i-o;
And on his farm he had a cow,
e-i-e-i-o;
With a moo-moo here and a moo-moo-there
Here a moo-there a moo-everywhere a moo, moo.
Old McDonald had a farm e-i-e-i-o.
Repeat. Change animals to:
**pig**—oink-oink
**duck**—quack-quack
**horse**—neigh-neigh
**donkey**—hee-haw
**chicken**—chick-chick
You may add your own animals, as well.

 

# Answer Key

**page 7**

1. corn
2. apple
3. cow
4. hen
5. tractor
6. barn

**page 12**

hen-eggs
pig-ham
cow-milk
corn-cereal
pumpkin-pie

**page 22**

I live on a farm. There are *c*ows, *sh*eep, *h*ens, and many *h*orses. We all have jobs to do. *M*other milks the cow. I feed the *ch*ickens. The animals live in a *b*arn. It is fun to live on a farm.

**page 60**

1. big
2. little
3. in
4. out
5. under
6. over
7. on
8. off

**page 61**

dog
pig
duck
sheep
hen
goat
rabbit

**page 62**

CORN
EGG
WHEAT
PUMPKIN
BEANS
MILK
HAM

**page 66**

1. boot
2. egg
3. cabbage
4. seed

**page 67**

$1 + 2 = 3$
$2 + 2 = 4$
$3 + 2 = 5$
$1 + 1 = 2$
$3 + 3 = 6$
$3 + 3 = 6$
$2 + 3 = 5$
$4 + 1 = 5$

**page 68**

$3 + 4 = 7$
$2 + 3 = 5$
$4 + 2 = 6$
$3 + 5 = 8$
$2 + 2 = 4$
$4 + 5 = 9$

**page 72**

1. milk-cheese
2. yarn-sweater
3. bone-meat
4. apple-pear
5. hat-boot
6. flowers-seeds

# Resources

**Book**

Aylesworth, Jim. *One Crow*. Lippincott, 1988.

Brown, Margaret Wise. *Big Red Barn*. Harper Collins, 1989.

Clements, Andrew. *Who Owns the Cow?* Clarion, 1995.

Cook, Brenda. *All About Farm Animals*. Doubleday, 1988.

Ehlert, Lois. *Color Farm*. Harper Collins, 1990.
*Growing Vegetable Soup*. HBJ, 1987.

Fleming, Denise. *Barnyard Banter*. Holt, 1994.

Florian, Douglas. *Vegetable Garden*. HBJ, 1991.

Fox, Mem. *Hattie and the Fox*. Macmillan, 1987.

Garland, Michael. *My Cousin Katie*. Crowell, 1989.

Gibbons, Gail. *Farming*. Holiday House, 1988.

Gibbons, Gail. *The Milkmakers*. Macmillan, 1985.

Henderson, Kathy. *I Can Be a Farmer*. Children's Press, 1989.

Henley, Claire. *Farm Day*. Dial, 1991.

Hill, Eric. *Spot Goes to the Farm*. Putnam, 1987.

Jackson, Woody. *Counting Cows*. HBJ, 1995.

Kightley, Rosalinda. *The Farmer*. Macmillan, 1988.

Kitchen, Bert. *Pig in a Barrow*. Dial, 1991.

Lewison, Wendy Cheyette. *Going to Sleep on the Farm*. Dial, 1992.

Manson, Christopher. *A Farmyard Song*. North-South Books, 1992.

Miller, Jane. *Farm Alphabet Book*. Scholastic, 1981.

Miller, Jane. *Farm Noises*. Simon and Schuster, 1989.

Most, Bernard. *The Cow That Went Oink*. Harcourt Brace, 1990.

Pearce, Q. L. and W. J. Pearce. *In the Barnyard*. Silver Press, 1990.

Provensen, Alice and Martin. *Our Animal Friends at Maple Hill Farm*. Random House, 1974.

Provensen, Alice and Martin. *The Year at Maple Hill Farm*. Atheneum, 1978.

Sendak, Maurice. *Chicken Soup and Rice*. Collins, 1962.

Siebert, Diane. *Heartland*. Crowell, 1989.

Tafuri, Nancy. *This Is a Farmer*. Greenwillow, 1994.

Thompson, Graham. *Tractors*. Gareth Stevens Publ., 1986.

Tresselt, Alvin. *Wake Up, Farm!* Lothrop, Lee, Shepard, 1991.

Wildsmith, Brian and Rebecca. *Wake Up, Wake Up!* HBJ, 1993.

**Technology CD-ROM**

*Animal Antics* CD, CD-Rom, MAC/WIN Hybrid CD, Jostens Home Learning.

*Farm Animals*, disk, MAC/WIN, CDL Software Shop.

*Farm Animals Close Up and Very Personal*, National School Products,VHS video, 1995.

*The Farm Visit*, disk, MAC/WIN, CDL Software Shop.

*Hop To It!*, disk, Apple II, Macintosh, Sunburst.

*In the Barn*, disk, MAC/WIN, CDL Software Shop.

*Junior Encyclopedia: The Farm* CD, CD-ROM, WIN CD, Humongous.

*Katie's Farm*, disk, IBM, MAC, CDL Software Shop.

*Learn About Plants*, disk, Macintosh, Apple IIe, IIc, IIGS, Sunburst.

*Math for Every Season*, National Geographic, filmstrip, 1985.

*Number Farm*, disk, IBM, Apple, Tandy, DLM.

*Seasons*, CD-ROM, MAC/WIN, National Geographic.

*Sitting on the Farm,* CD-ROM, MAC, MPC, Educational Resources, (ESL)

*Tip-Top Tots: The Nutrition Pyramid for Preschoolers*, National

*A World of Plants*, CD-ROM, Macintosh, Sunburst. School Products, VHS video, 1995.

 